SOLAR RADIATION GEOMETRY

By

Vanita Thakkar

(vanitaa.thakkar@gmail.com)

CONTENTS

SR. NO.	TITLE	PAGE NO.
1.	THE SOLAR ENERGY OPTION	3
2.	SOLAR ENERGY ON EARTH	5

- EARTH'S ROTATION
- EARTH'S REVOLUTION
    - Important positions in Earth's path of Revolution
    - Extra-terrestrial Energy Flux
        - Solar Constant
- TILT OF EARTH'S AXIS
    - Solstice And Equinox
- SPECTRAL DISTRIBUTION OF SOLAR RADIATION

- SOLAR RADIATION ON EARTH'S SURFACE

3. MEASUREMENT OF SOLAR RADIATION   21
   - PYRANOMETERS
   - PYRHELIOMETERS
   - SUNSHINE DURATION MEASUREMENTS

4. SOLAR ANGLES   35
   - BASIC ANGLES
     - Latitude-Longitude
     - Declination
     - Hour Angle
     - Local Solar Time
     - Equation of Time
   - DERIVED ANGLES
     - Related to relative position of Sun
     - Related to the orientation of surface intercepting solar radiation
   - RELATIONS BETWEEN BASIC AND DERIVED ANGLES
   - GENERAL EQUATION FOR ANGLE OF INCIDENCE
   - DAY LENGTH

THE SOLAR ENERGY OPTION

- **Sun** is the basic source of Energy for Earth.
- Solar Energy is available in the form of **Electromagnetic Radiations**.
- **Sun** is a large sphere of very hot gases, heat being generated by the various fusion reactions in it.
- **Diameter** of **Sun** = 1.39×10^6 km.
- **Diameter** of **Earth** = 1.27×10^4 km.
- Sun **subtends an angle** of only **32'** at **Earth's surface** (because of large distance between them).
- The direct / beam radiation received from the Sun on the earth is almost parallel.
- Brightness of the Sun varies from its centre to its edge, however, for engineering calculations, it is assumed to be uniform over the entire solar disc.
- The sun generates an enormous amount of energy - approximately **1.1×10^{20} kilowatt-hours every second. (A kilowatt-hour is the amount of energy needed to power a 100 watt light bulb for ten hours.)**

Figure – 2.1 : Sun – The Basic Energy Source.

SOLAR ENERGY ON EARTH

- The earth's outer atmosphere intercepts about **one two-billionth** of the energy generated by the sun, or about **1500 quadrillion (1.5×10^{18}) kilowatt-hours per year.**
- Due to **Reflection, Scattering, and Absorption** by **gases and aerosols** in the atmosphere, however, only **47%** of this, or approximately **700 quadrillion (7×10^{17}) kilowatt-hours per year**, reaches the surface of the earth.
- In the earth's atmosphere, solar radiation is received :
  - directly **(Direct Radiation).**
  - by diffusion in air, dust, water, etc., contained in the atmosphere **(Diffuse Radiation).**
- **Global radiation = Direct Radiation + Diffuse Radiation.**
- The amount of **incident energy per unit area and day** depends on a number of factors, e.g. :
  - Solar Radiation Geometry, which includes Solar Angles, Locational or Geographical factors and Season / Time of the year.
  - Local climate.
  - Inclination of the collecting surface in the direction of the sun.

Earth's Rotation :

- The term **Earth's rotation** refers to the **spinning** of Earth on **its axis**.
- Due to rotation, the Earth's surface moves at the equator at a speed of about **467 m per second** or **slightly over 1675 km per hour**.
- One rotation takes exactly **twenty-four hours** and is called a **mean solar day**.
- The Earth's rotation is responsible for the **daily cycles of day and night** – i.e. at any moment in time, one half of the Earth is in sunlight, while the other half is in darkness.
- The edge dividing the daylight from night is called the **circle of illumination**, which is a **line** that **bisects areas on the Earth receiving sunlight** and **those areas in darkness**. It cuts the **spherical Earth** into **lighted** and **dark halves**.
- The Earth's rotation also creates the **apparent movement of the Sun across the horizon**.

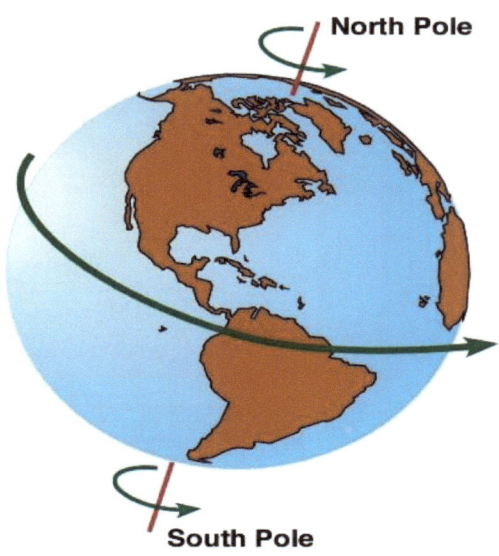

Figure – 2.2 : Earth's Rotation.

- Looking down at the **Earth's North Pole** from space one would see that the direction of rotation is **counter-clockwise.**
- Looking down at the **Earth's South Pole** from space one would see that the direction of rotation is **clockwise.**

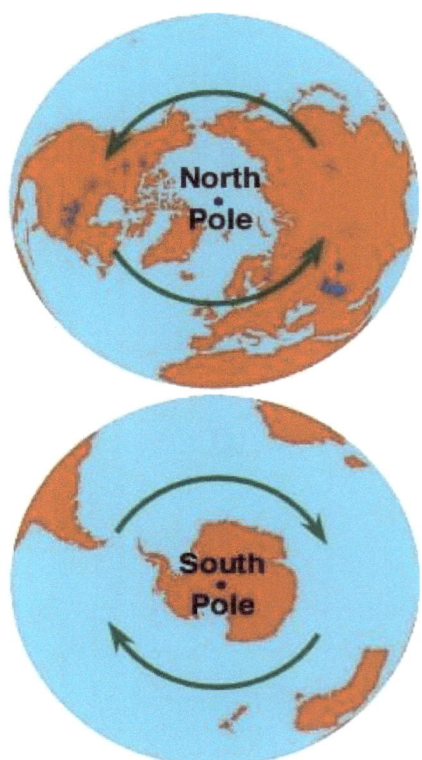

Figure – 2.3 : Direction of Earth's Rotation.

Earth's Revolution :

- The orbit of the Earth around the Sun is called an **Earth revolution**.
- This celestial motion takes **365.26** days to complete one cycle.
- Earth's orbit around the Sun is not circular, but oval or **elliptical**.

- An elliptical orbit causes the Earth's distance from the Sun to vary over a year. Yet, this phenomenon is **not responsible for the Earth's seasons** !
- This variation in the distance from the Sun causes the amount of solar radiation received by the Earth to annually vary by about **6%**.

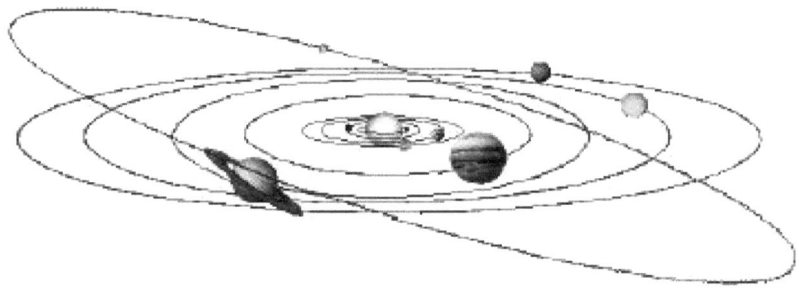

Figure - 2.4 : Earth's Revolution.

Important positions in Earth's path of Revolution :

- On **January 3, Perihelion**, Earth is **closest** to the Sun **(147.3 million km)**.
- On **July 4, Aphelion**, Earth is **farthest** from the Sun **(152.1 million km)**.
- **Average distance** of **Earth** from the **Sun** over a period of **one-year** is ~ **149.6 million km, i. e. 1.496 × 10⁸ km**.
- The interesting observation here is that in the northern hemisphere, January 3, i.e. Perihelion, when

the Earth is closest to the Sun is during winter, so a cold day and July 4, i.e. Aphelion, when the Earth is farthest from the Sun is during Summer, so it is a hot day. This is due to the fact that seasons created on the Earth are more affected by its tilt about its axis than its distance from the Sun.

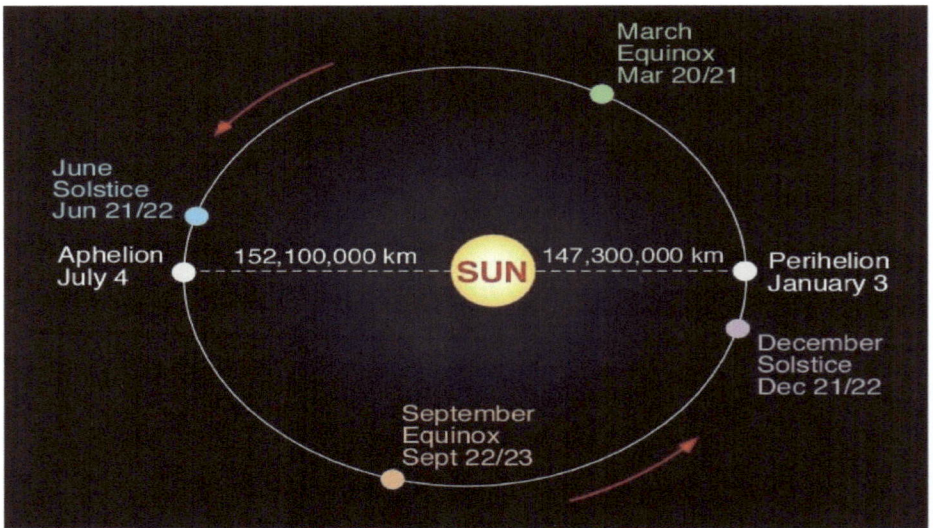

Figure – 2.5 : Important Positions on Earth's path of Revolution.

Extra-terrestrial Energy Flux :

- The **Energy Flux** (rate of energy transfer through a unit area) received from Sun **outside the Earth's atmosphere** is called **Extra-terrestrial Energy Flux**.

SOLAR RADIATION GEOMETRY

- Measurements indicate that the **Extra-terrestrial Energy Flux** received from Sun **over the Earth's atmosphere** is essentially constant, which is called the Solar Constant and which can be defined as :

SOLAR CONSTANT (I_{SC}) : The <u>rate</u> at which **Energy is received** from the Sun on a **unit area perpendicular to the rays of Sun outside the earth's atmosphere**, at a **mean distance** of the Earth from the Sun (~ 1.496×10^8 km) is called Solar Constant and it is given by :

$$I_{SC} = 1367 \text{ W/m}^2$$

The solar constant includes all types of solar radiation, not just the visible light.

Variations in Extra-terrestrial Energy Flux : Due to variation in Earth-Sun distance throughout the year, the Extra-terrestrial Flux varies, which can be calculated from the equation :

$$I_{SC}' = I_{SC} [1 + 0.033 \cos (360n/365)],$$

Where, n = the number of day of the year.

Tilt of Earth's Axis :

- The **ecliptic plane** can be defined as a **two-dimensional flat surface** that geometrically intersects the Earth's orbital path around the Sun.

SOLAR RADIATION GEOMETRY

- On this plane, the **Earth's axis** is **not at right angles** to this surface, but **inclined** at an **angle of about 23.5°** from the perpendicular.

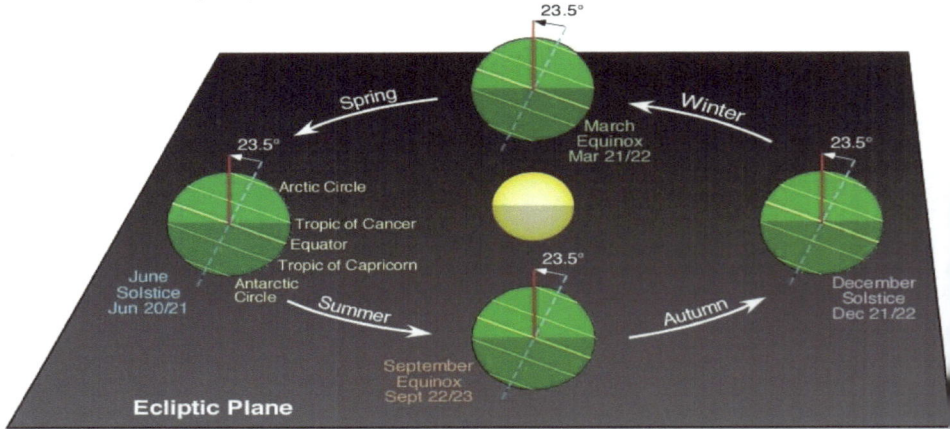

Figure – 2.6 : Equinox and Solstice.

Solstice and Equinox :
- Figure 2.6 shows a side view of the Earth in its orbit about the Sun on four important dates: June solstice, September equinox, December solstice, and March equinox.
- Angle of the Earth's axis in relation to the Ecliptic Plane and the North Star on these four dates remains unchanged.
- Yet, the **relative position** of the **Earth's axis to the Sun does change** during this cycle.
- This is responsible for the **annual changes in the height of the Sun** above the **horizon**.

- It also causes the **seasons**, by controlling the intensity and duration of sunlight received by locations on the Earth.
- In an overhead view of the same phenomenon, one can see how the **circle of illumination changes its position** on the Earth's surface, as shown in figure 2.7.
- During the **two equinoxes**, the circle of illumination **cuts through North Pole and South Pole**.
- On the **June solstice**, the **circle of illumination** is **tangent to the Arctic Circle (66.5° N)** and the region above this latitude receives **24 hours of daylight**. The Arctic Circle is in **24 hours of darkness** during the **December solstice**.

SOLAR RADIATION GEOMETRY

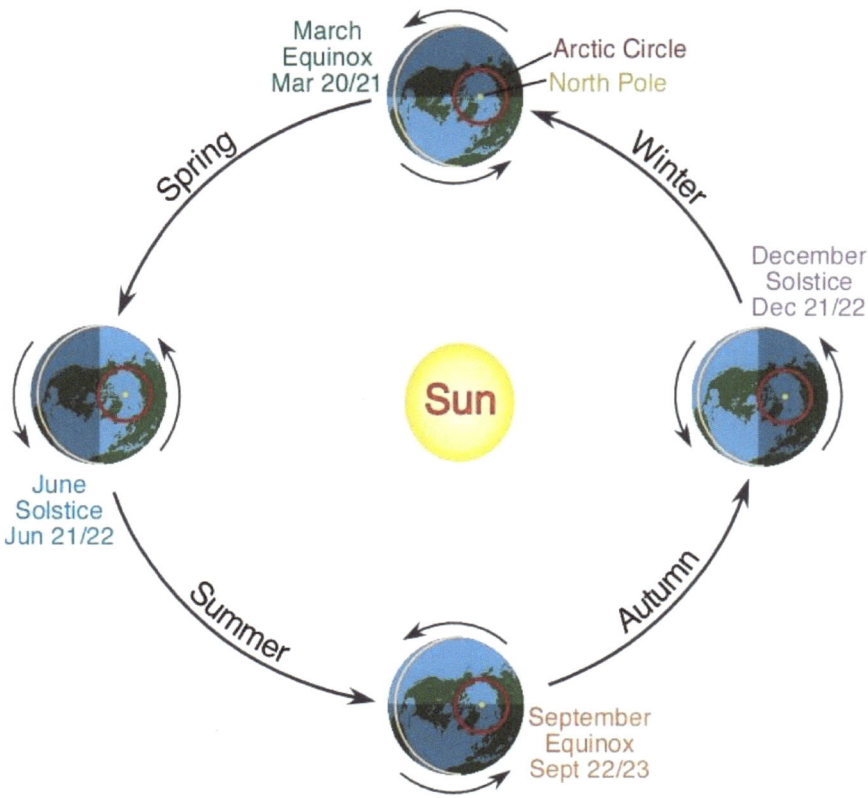

Figure – 2.7 : Equinox and Solstice – overhead view.

SOLSTICE :

- **On June 21 or 22** (also called the **summer solstice** in Northern Hemisphere) Earth is positioned in its orbit so that **North Pole is leaning 23.5° toward the Sun** : All locations **north of the equator** have **day lengths greater than twelve hours**, while all locations **south of the equator** have **day lengths less than twelve hours**.

- **On December 21 or 22** (also called the **winter solstice** in Northern Hemisphere), Earth is positioned so that **South Pole is leaning 23.5 degrees toward the Sun** : All locations **north of the equator** have **day lengths less than twelve hours**, while all locations **south of the equator** have day lengths exceeding twelve hours.

Figure – 2.8 : Solstice.

EQUINOX :

- **On September 22 or 23**, also called the **autumnal equinox** in the Northern Hemisphere, neither pole is tilted toward or away from the Sun.
- In the Northern Hemisphere, **March 20 or 21** marks the arrival of the **vernal equinox** or **spring** when once again the poles are not tilted toward or away from the Sun.
- **Day lengths** on both of these days, **regardless of latitude, are exactly 12 hours.**

SOLAR RADIATION GEOMETRY

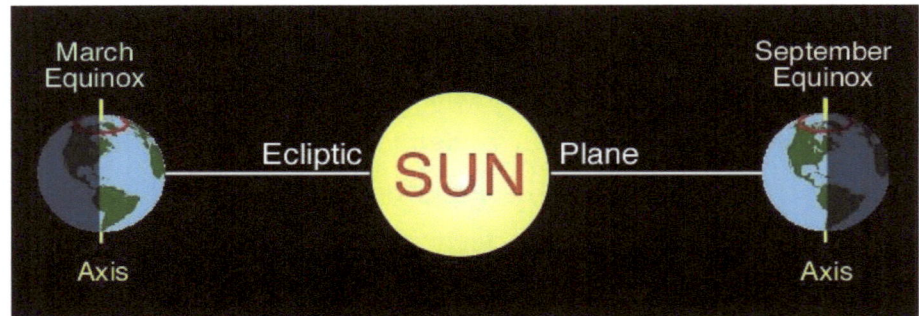

Figure - 2.9 : Equinox.

Spectral Distribution of Solar Radiation :

- **Solar Radiation spectrum** is close to that of a **black body** with a temperature of **about 5,800 K.**
- About **half** that lies in the **visible short-wave part** of it and the other half mostly in **near-infrared part.**
- Some also lies in the **ultraviolet part** of the spectrum.
- **Spectrum** of **electromagnetic radiation** striking Earth's atmosphere : **100 to 10^6 nm,** can be divided into **five regions,** as shown in figure 2.10 :
    - **Ultraviolet C (UVC) : 100 to 280 nm.** Radiation Frequency > violet light Frequency (so, **invisible** to human eye). Mostly absorbed by **Lithosphere.**
    - **Ultraviolet B (UVB) : 280 to 315 nm. Mostly absorbed by atmosphere;** Along with UVC, responsible for **photochemical reaction** leading to production of **Ozone layer.**

- Ultraviolet A (UVA) : 315 to 400 nm. Considered less damaging to the DNA.
- Visible range or light : 400 to 700 nm.
- Infrared range : 700 nm to 106 nm. An important part of the electromagnetic radiation reaching Earth. Divided into three types :
  - Infrared-A: 700 nm to 1,400 nm.
  - Infrared-B: 1,400 nm to 3,000 nm.
  - Infrared-C: 3,000 nm to 1 mm.

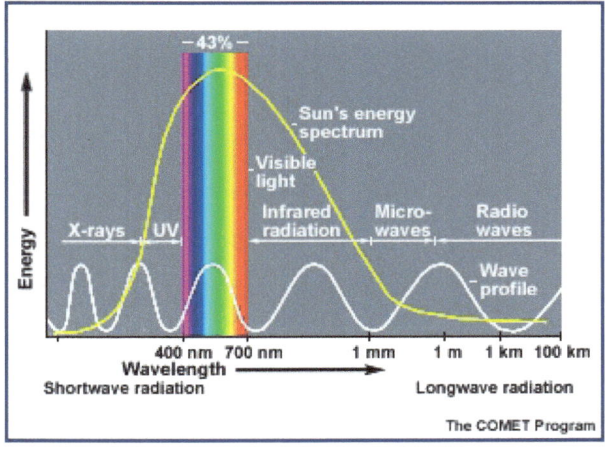

Figure – 2.10 : Spectral Distribution of Solar Radiation.

- Maximum value of Solar Radiation Intensity = 2074 W/m2 occurs at 0.48mm wavelength.
- 99% of solar radiation is obtained upto a wavelength of 4mm.
- Table 2.1 shows the solar energy flux obtained in different ranges of wavelengths :

Table 2.1 : Solar Energy Flux obtained in different ranges of wavelengths

Wave length (μm)	0.00-0.38	0.38-0.78	0.78-4.0
Approx. energy (W/m^2)	95	640	618
Approx. % of total energy	7%	47.3%	45.7%

Solar Radiation on Earth's Surface :

- Solar radiation reaching Earth's surface **differs** in **amount and character** from Extraterrestrial Radiation. It attenuates, as shown in figure 2.11 :
  - Part of the radiation is **reflected back**, especially by **clouds.**
  - Some part of it is **absorbed** by **molecules in air** such as –
    - O_2 and O_3 **(ozone)** absorb nearly all **UV radiations.**
    - **Water vapour and CO_2 absorb** some energy in **infrared range.**
  - Some part of it is **scattered** by **droplets** in **clouds and dust particles.**

SOLAR RADIATION GEOMETRY

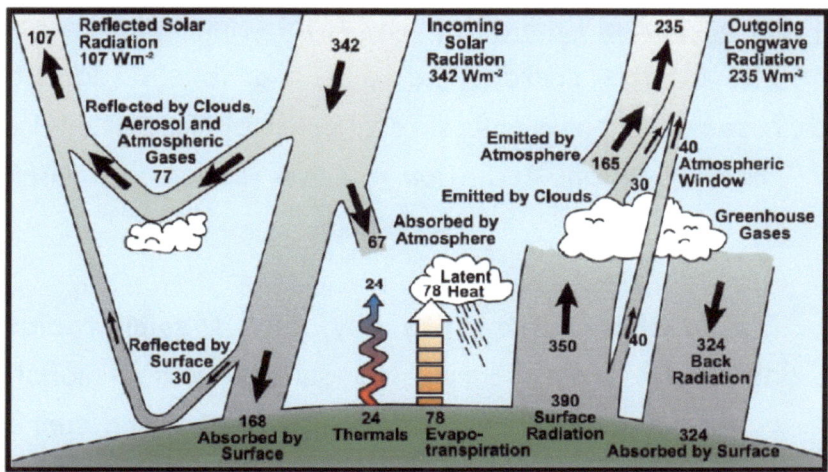

Figure – 2.11 : **Attenuation of Solar Radiation.**

- Attenuation of solar radiation results in the incidence of radiation on the Earth's surface in two different forms, viz. direct or beam radiation and diffuse radiation, the combination of which is called total or global radiation.

Direct / Beam Radiation : Solar radiation that does not get absorbed or scattered, but **reaches the ground directly** from the Sun. It produces shadow when interrupted by an opaque object.

Diffuse Radiation : Solar radiation received after its **direction** has been **changed** by **reflection and scattering** in the atmosphere.

Total or Global Radiation: The total solar radiation received on the Earth's surface, i.e. direct or beam radiation and diffuse radiation, is called Total or Global Radiation, i.e.

Total or Global Radiation = Beam Radiation + Diffuse Radiation.

- During the course of a day, direct solar radiation is observed from sunrise to sunset, while global solar radiation is observed in the twilight before sunrise and after sunset, despite its diminished intensity at these times.
- Direct solar radiation is prominent on clear / cloudless days, while diffuse solar radiation is predominant during cloudy, overcast days.

MEASUREMENT OF SOLAR RADIATION

- Solar energy on Earth is measured by instruments, broadly known as **Actiometers or Solar Radiometers**, which measure **heating power of radiation**.
- An actinometer or a solar radiometer **absorbs solar radiation** at its sensor, **transforms** it into **heat** and **measures** the resulting amount of **heat** to ascertain the level of solar radiation.
- Methods of measuring heat include taking out heat flux :
    - As a **temperature change** e.g. in water flow pyrheliometer, silver-disk pyrheliometer or bimetallic pyranograph.
    - As a **thermoelectromotive** force e.g., in thermoelectric pyrheliometer or thermoelectric pyranometer.
- These days, types of radiometers / actinometers using a thermopile are generally used.
- The radiometers used for ordinary observation are Pyrheliometers, which measure direct solar radiation and Pyranometers, which measure global solar radiation.

Pyranometer :
- A pyranometer is used to measure **global solar radiation** falling on a **horizontal surface**.
- Its sensor has a **horizontal radiation-sensing surface** that **absorbs solar radiation energy** from the **whole sky**

(i.e. a solid angle of 2π sr) and transforms this energy into **heat**. Global solar radiation can be ascertained by measuring this heat energy.

- Most pyranometers in general use are now the **thermopile type**, although **bimetallic pyranographs / pyranometers** are occasionally found.
- Thermoelectric pyranometers and bimetallic pyranometers / pyranographs are both **hermetically sealed** in a glass dome to protect the sensor portion from wind and rain and prevent the sensor surface temperature from being disturbed by wind.
- A **desiccant** is placed in the dome to **prevent condensation** from forming on the inner surface. The glass allows the passage of solar radiation in wavelengths from about 0.3 to 3.0 μm – a range that covers most of the sun's radiation energy.
- Some models are equipped with a **fan** to prevent dust or frost, which greatly affect the amount of light received, from collecting on the dome's outer surface.
- It is necessary to **check and clean** the glass surface at regular intervals to ensure that the dome wall constantly allows the passage of solar radiation.
- A **typical pyranometer** is schematically represented in Fig. 2.12 (a). It consists of a **white disk** for limiting the **acceptance angle** to **180°** and **two concentric hemispherical transparent covers made of glass**. The two

domes shield the sensor from thermal convection, protect it against weather threat (rain, wind, and dust) and limit the spectral sensitivity of the instrument in the wavelength range 0.3 to 3.0 μm. A **cartridge of silica gel** (desiccant) inside the dome absorbs water vapour.

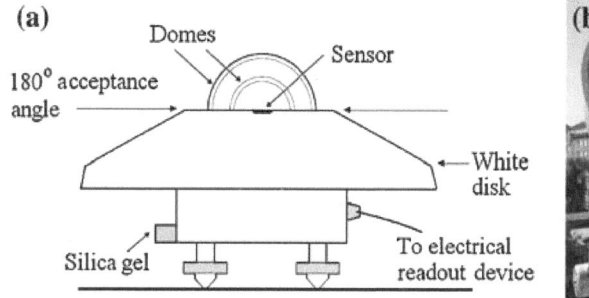 

Figure – 2.12 : (a) Pyranometer; (b) Pyranometer with shading ring to measure diffuse irradiance

- A pyranometer can be also used to measure the diffuse solar irradiance, often denoted by G_d or I_d, provided that the contribution of the direct beam component is eliminated.
- For this, a small shading disk can be mounted on an automated solar tracker to ensure that the pyranometer is continuously shaded.
- Alternatively, a shadow ring may prevent the direct component, often denoted by G_b or I_b, from reaching the sensor whole day long, as shown in Fig. 2.12 (b).

- As the daily maximum Sun elevation angle changes day by day, it is necessary to change periodically (days lag) the height of the shadow ring.

Thermoelectric Pyranometers : Thermoelectric Pyranometers, which use thermopiles are shown in figure 2.13. The principle of working of these pyranometers is as described below :

- Here, a **temperature difference** is produced between the sensor surface, called the **hot junction** and the reference temperature point, called the **cold junction**.
- As the **temperature difference** is **proportional** to the intensity of the radiation absorbed, the level of solar radiation can be derived by measuring the **thermoelectromotive force** from the thermopile.
- Since this type of pyranometer is a relative instrument, **calibration** should be performed to determine the **instrumental factor** through comparison with a standard instrument. As the thermoelectromotive force output depends on the **unit's temperature**, the temperature inside the instrument enclosure should be monitored to enable correction.

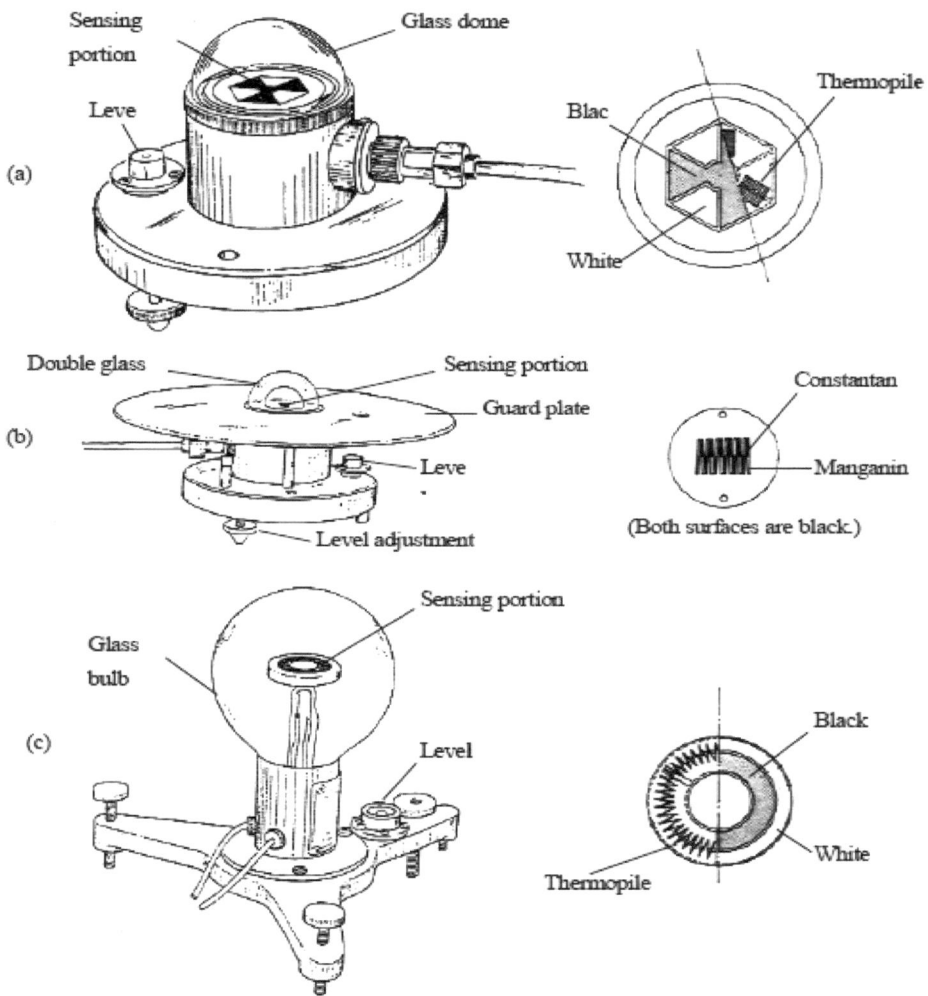

Figure – 2.13 : Thermoelectric Pyranometers and their sensing elements.

Bimetallic Pyranograph :

- A bimetallic pyranograph is shown in Figure 2.14.
- The **radiation-sensing element** (in the upper right of the figure) consists of **two pairs of bimetals, one painted black** and the **other painted white**, placed in **opposite directions** (face up and face down) and attached to a **common metal plate** at **one end**.
- At the **other end**, the **white bimetallic strips** are fixed to the frame of the pyranograph, and the **black ones** are connected to the **recorder section** via a **transmission shaft**.
- The **deflection** of the **free edge** of the black strips is transmitted to the **recording pen** through a **magnifying system**. When the **air temperature changes**, the **black and white strips** attached to the **common plate** at one end both **bend** by the same amount but in **opposite directions**. As a result, only the **temperature difference** attributed to solar radiation is transmitted to the recording pen.

SOLAR RADIATION GEOMETRY

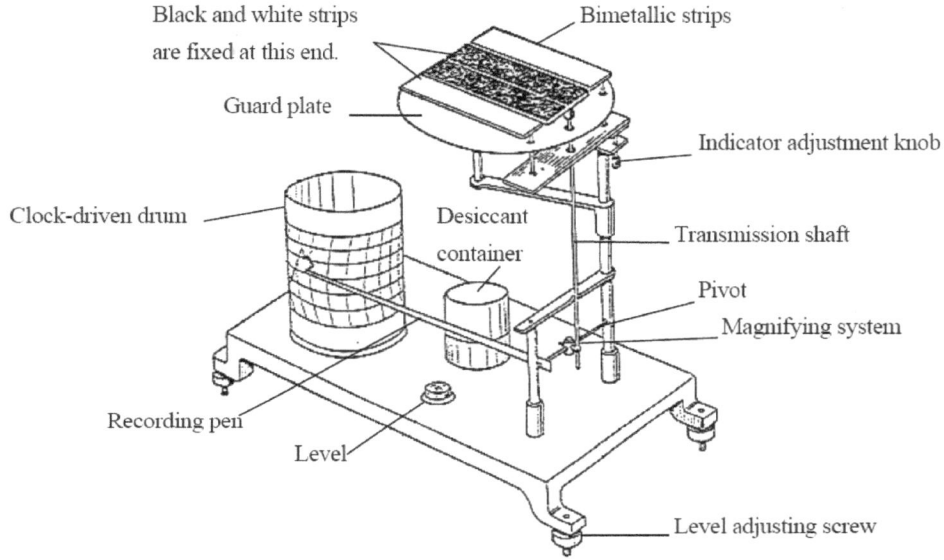

Figure - 2.14 : Bimetallic Pyranograph

Pyrheliometer :

- A pyrheliometer is used to **measure direct solar radiation** from the sun and its marginal periphery.
- To measure direct solar radiation correctly, its **receiving surface** must be arranged to be **normal** to the **solar direction**. For this reason, the instrument is usually mounted on a **sun-tracking device** called an **equatorial mount**. A **two-axis Sun tracking mechanism**, most often used for sun-tracking.
- Different types of Pyrheliometers are described as follows :

Thermoelectric Pyrheliometer :

- The structure of a thermoelectric pyrheliometer is shown in Figure 2.15. This instrument uses a **thermopile** at its *sensor*, and continuously delivers a thermoelectromotive force in proportion to the direct solar irradiance.
- The pyrheliometer consists of a **detector**, which is a **multi-junction thermopile** placed at the bottom of a **collimating tube**, provided with a **quartz window** to protect the instrument, as shown in the schematic diagram of a typical pyrheliometer in figure 2.15.
- **Copper-plated constantan** wire is used as the thermopile in the sensor portion, which is attached to the **bottom of the collimator tube / cylinder** at **right angles** to its axis.
- The cylinder is fitted with **diaphragms** to direct sunlight to the sensor portion. It is made of a **metallic block** with **high heat capacity** and **good thermal conductivity**, and is enclosed in a **polished intermediate cylinder** and a **silver-plated outer brass cylinder** with **high reflectivity** to prevent rapid ambient temperature changes or outer wind from disturbing the heat flux in the radiation-sensing element. The cylinder is kept dry using a **desiccant** to prevent condensation on the inside of the aperture window.

- The **detector is coated** with **optical black paint** (acting as a **full absorber** for solar energy in the wavelengths range 0.280-3 mm). Its **temperature** is **compensated** to minimize sensitivity of ambient temperature fluctuations.
- The pyrheliometer **aperture angle** is **5°**. Consequently, radiation is received from the Sun and a **limited circumsolar region**, but **all diffuse radiation** from the rest of the sky is **excluded**.
- A **readout device** is used to give the instant value of the direct beam irradiance. Its **scale** is adapted to the sensitivity of the particular instrument in order to display the value in SI units, **W / m^2**.
- Figure 2.16 shows the **Hukseflux DR01 first class pyrheliometer**.

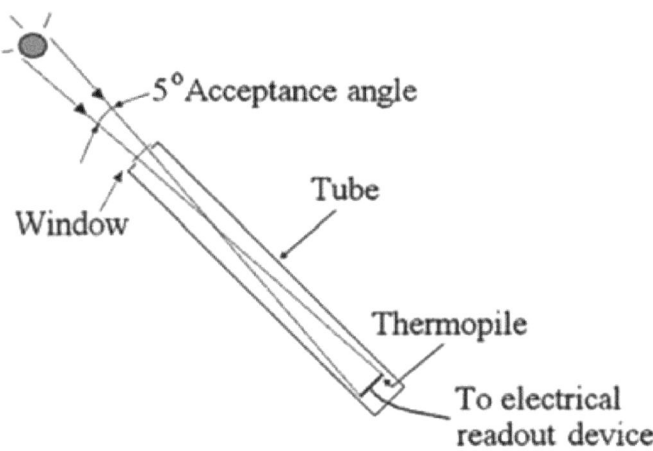

Figure - 2.15 : Schematic Diagram of Pyrheliometer

SOLAR RADIATION GEOMETRY

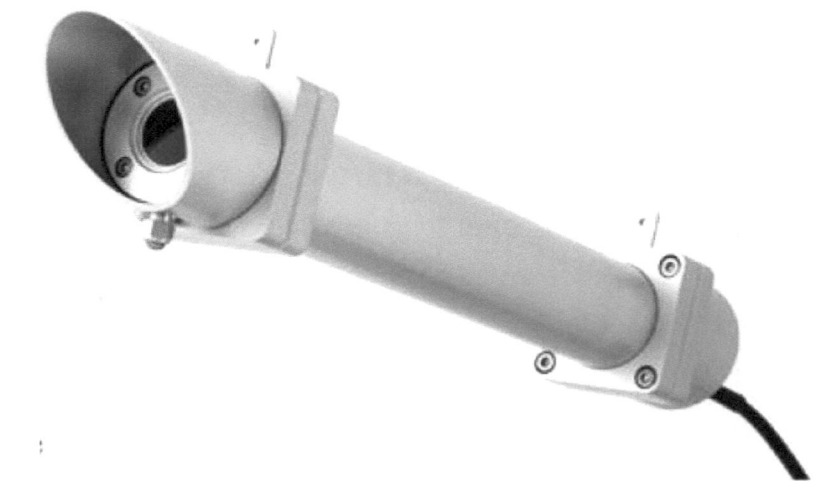

Figure – 2.16 : Hukseflux DR01 first class pyrheliometer

Sunshine Duration Measurements : According to **World Meteorological Organization (WMO)**, sunshine duration in a given period is defined as the sum of the **time intervals** for which the **direct solar irradiance exceeds** the threshold of **120 W/m²**. In practice, **two methods** are widely used for measuring sunshine duration—**burning card method** and **pyranometric method**, out of which burning card method is more widely adopted. Pyranometric method uses pyranometer and calculative analytical procedure involving use of statistical historical data, hence is a complicated and at times, less accurate method of finding out sunshine duration.

Burning card method is based on the **Campbell-Stokes sunshine recorder**, which basic setup consists of a **glass sphere mounted concentrically** in a **segment of a spherical bowl**, as shown in figure 2.19. The **support** is **adjustable** so that the **axis of the sphere** may be **inclined** to the angle of the **local latitude**. The **spherical bowl segment** holds the **recording card**. The **glass sphere focuses** the **direct beam solar radiation** on to the **card**, **burning** a trace whenever the Sun is shining. The **position and length** of the **trace** indicate the starting time and duration of the sunshine interval.

The errors of this recorder are mainly due to the dependence of **burning initiation** on **card's temperature and humidity** as well as to the **over-burning effect**, especially in case of **broken clouds**.

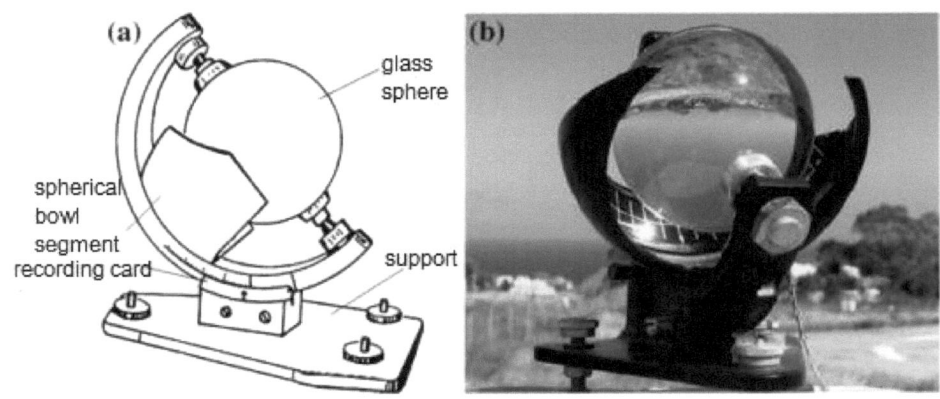

Figure – 2.19 : (a) Schematic Diagram of Campbell-Stokes Sunshine Recorder; (b) Photo of a typical Campbell-Stokes sunshine recorder.

Different types of Sunshine Recorders are described below :
Campbell-Stokes Sunshine Recorder : A Campbell-Stokes sunshine recorder **concentrates sunlight** through a **glass sphere**

onto a **recording card** placed at its **focal point**. The **length of the burn trace** left on the card represents the **sunshine duration**. It is the most commonly used Sunshine Recorder.

The device's structure is shown in Figure 2.20 (a). Its structure and principle of working are as follows :

- A **homogeneous transparent glass sphere** L is supported on an arc XY, and is focused so that an image of the sun is formed on recording paper placed in a **metal bowl** FF' attached to the arc.
- The **glass sphere** is **concentric** to this **bowl**, which has **three partially overlapping grooves** into which **recording cards** for use in the **summer, winter or spring and autumn** are set, as shown in Figure 2.20 (b). Three different recording cards, as shown in Figure 2.20 (c) are used depending on the season.
- The focus shifts as the sun moves, and a **burn trace** is left on the **recording card** at the focal point. A burn trace at a particular point indicates the presence of sunshine at that time, and the **recording card** is **scaled with hour marks** so that the exact time of sunshine occurrence can be ascertained. Measuring the overall length of burn traces reveals the sunshine duration for that day.
- For exact measurement, the sunshine recorder must be **accurately adjusted** for **planar leveling, meridional direction and latitude**.
- Campbell-Stokes as well as Jordan sunshine recorders mark the occurrence of sunshine on recording paper at a position

corresponding to the **azimuth of the sun** at the site, and the time of sunshine occurrence is expressed in **local apparent time**.

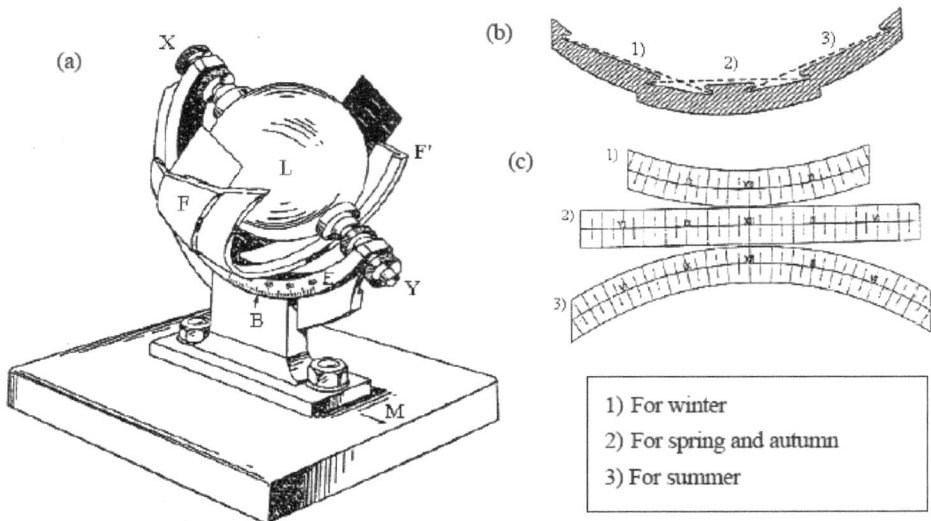

Figure – 2.20 : Campbell-Stokes Sunshine Recorder; (a) Structure; (b) Cross section of Bowl and Grove; (c) Recording Cards.

To obtain uniform results for observation of sunshine duration with a Campbell-Stokes sunshine recorder, the following points should be noted when reading records:

- If the **burn trace** is **distinct and rounded at the ends**, subtract half of the curvature radius of the trace's ends from the trace length at both ends. Usually, this is equivalent to subtracting 0.1 hours from the length of each burn trace.

- If the burn trace has a **circular form**, take the radius as its length. If there are multiple circular burns, count two or three as a sunshine duration of 0.1 hours, and four, five or six as 0.2 hours. Count sunshine duration this way in increments of 0.1 hours.
- If the burn trace is **narrow**, or if the recording card is only slightly discolored, measure its entire length.
- If a **distinct burn trace diminishes in width by a third or more**, subtract 0.1 hours from the entire length for each place of diminishing width. However, the subtraction should not exceed half the total length of the burn trace.

SOLAR ANGLES

The **position** of a **point P** on the **Earth's surface** with respect to the **sun's rays** depends on the **basic solar angles of the location**, viz. the **latitude** of the location, ϕ, **hour angle**, ω, for the point, and the sun's **declination angle**, δ of the location for the day of the year under consideration. Figure 2.24 shows these three fundamental / basic angles. Point **P** represents a **location** on the **northern hemisphere**. Several other angles are useful in solar radiation calculations. Such angles include those **angles** which are **related to the relative positions of the Sun** during the course of a year, such as the **sun's zenith angle** θ_z, **altitude angle** α, and **azimuth angle** γ_s. For a particular **surface orientation**, the sun's **incidence angle** θ, and **surface-solar azimuth angle** γ, and the **slope angle s**, may be defined. All of these additional angles may be expressed in terms of the three basic angles.

Thus, angles useful in Solar Radiation Analysis can be listed as:

Basic Angles :
1. Latitude of location (ϕ).
2. Declination (δ).
3. Hour Angle (ω).

Derived Angles :
Related to the **relative position of the Sun** :

4. Altitude Angle (α).
5. Zenith Angle (θ_z).
6. Solar Azimuth angle (γ_s).

Related to the **orientation of surface intercepting solar radiation** :

7. Angle of Incidence (θ)
8. Surface-solar azimuth angle (γ)
9. Slope (s).

BASIC ANGLES :

Latitude of location (ϕ) : Angle made by **radial line joining the location to the centre of the Earth** and the **projection of that line on the Equatorial plane.**

It is the **angular distance north or south** of **equator** measured from **centre of Earth**.

It varies from **0° at equator** to **90° at the poles**.

Latitude angle is shown as ϕ in fig. 2.24. The intersection of latitude and longitude gives the position of a location on the Earth's surface, as shown in Figure 2.25.

SOLAR RADIATION GEOMETRY

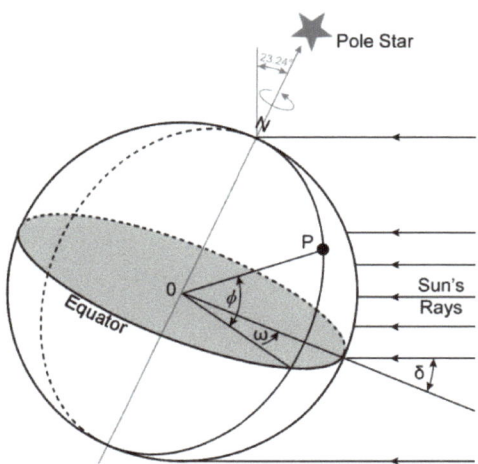

Figure – 2.24 : Basic Solar Angles - Latitude of location (φ), Declination (δ), Hour Angle (ω)

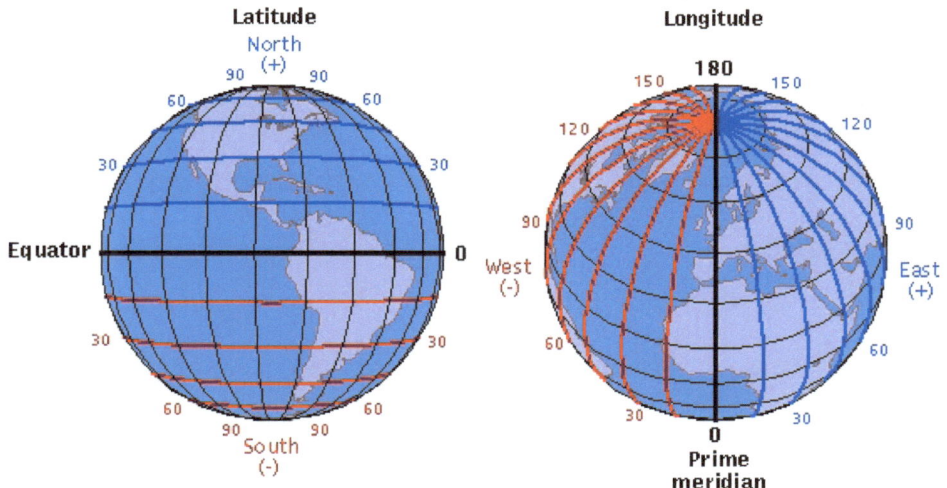

Figure – 2.25 : Latitude along with the longitude indicates the position of any point on earth.

SOLAR RADIATION GEOMETRY

Declination (δ) : Angular distance of the sun's rays north or south of the equator.

It is the **angle between** a <u>line extending from the centre of Sun to centre of Earth</u> and <u>the projection of this line upon Earth's Equatorial Plane</u>.

It is due to **tilt of Earth's axis** and it **varies between 23.5°** (Summer Solstice : June 22) to **-23.5°** (Winter Solstice : December 22).

On Equinoxes, Declination = 0.

Declination is given by :

$$\delta = 23.45° \cdot \sin\left[\frac{360°}{365} \cdot (N + 284)\right]$$

where, N = number corresponding to the day of the year.

Figure 2.26 shows declination on Equinoxes and Solstices and Figures 2.27 and 2.28 show variations in declination throughout the year and along the Earth's orbit, respectively.

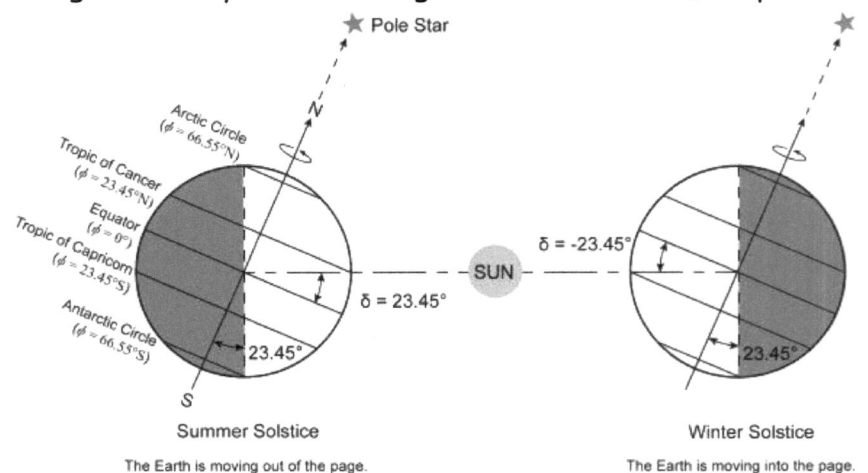

Summer Solstice
The Earth is moving out of the page.

Winter Solstice
The Earth is moving into the page.

SOLAR RADIATION GEOMETRY

Figure – 2.26 : Declination on Equinoxes and Solstices.

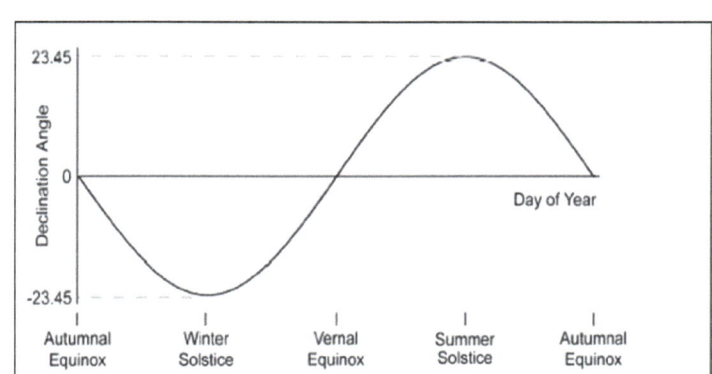

Figure – 2.27 : Variation in declination throughout the year.

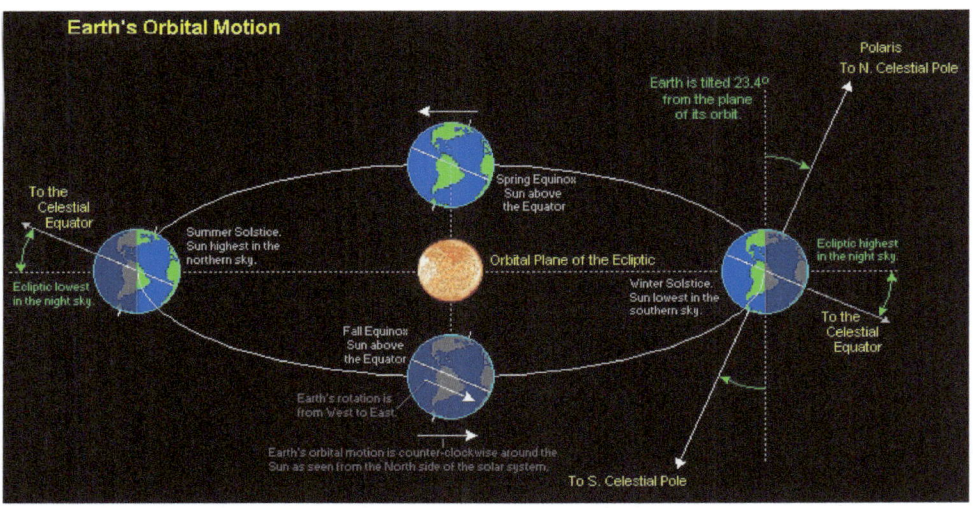

Figure – 2.28 : Variation in declination along the Earth's orbit.

Hour Angle (ω) : The angle through which the **earth must turn to bring the Meridian of a point directly in line with the Sun's rays**. It is **a measure of the time of the day** with respect to

SOLAR RADIATION GEOMETRY

solar noon (Solar noon occurs when **the sun** is at the **highest point in the sky** & ω **is symmetric** with respect to **solar noon**).

At noon, ω = 0.

Also, ω = 15° per hour.

The hour angle is **measured from noon**, based on the **Local Solar Time (LST)**, +ve <u>before noon</u> and -ve <u>during afternoon</u>.

Figure 2.29 shows hour angle and its variation during a day.

<u>LOCAL SOLAR TIME :</u> There is variation in local standard time compared local solar time owing to the longitudinal distance between the location under consideration and the location of the standard meridian. Local **Solar Time** can be obtained from Standard Time observed on a clock by applying **Two Corrections** :

1. Due to **difference in Longitude** between a **location** and **the meridian on which the standard Time is based** : Has a magnitude of **4 minutes** for **every degree difference in Longitude**.

2. Due to **Equation of Time** : The Earth's orbit and rate of rotation are subject to small perturbations, as the Earth's orbital velocity varies throughout the year. So, the local solar time as measured by a sundial varies slightly from the mean time kept by a clock running at a uniform rate. A civil day is exactly equal to 24 hours, whereas a solar day is approximately equal to 24 hours. This variation is

represented by Equation of Time (EOT) and is available as average values for different months of the year. The EOT may be considered as constant for a given day. An approximate equation for calculating EOT given by Spencer (1971) is :

EOT = 0.2292(0.75 + 1.868CosN – 32.077SinN – 4.615Cos2N – 40.89Sin2N),

Where, N = (n-1)(360/365); n is the day of the year (counted from January 1st)

Figure 2.30 shows graphical representation of Equation of Time (EOT).

Hence,

LST = Standard Time ± 4 (Standard Time Longitude – Longitude of Location) + (Equation of Time Correction)

The **negative sign** is applicable for **Eastern Hemisphere.**

And

Hour Angle, ω = 15 (12 – LST)

Problem 1 : Determine Hour Angle for : 09:00 AM, 11:00 AM, 02:00 PM, 04:30 PM.

Solution :

Hour Angle, ω = 15 (12 - LST)

09:00 AM : ω = 15(12 - 9) = 45°

11:00 AM : ω = 15(12 - 11) = 15°

02:00 PM : ω = 15(12 - 14) = -30°

04:30 AM : ω = 15(12 - 16.5) = -67.5°

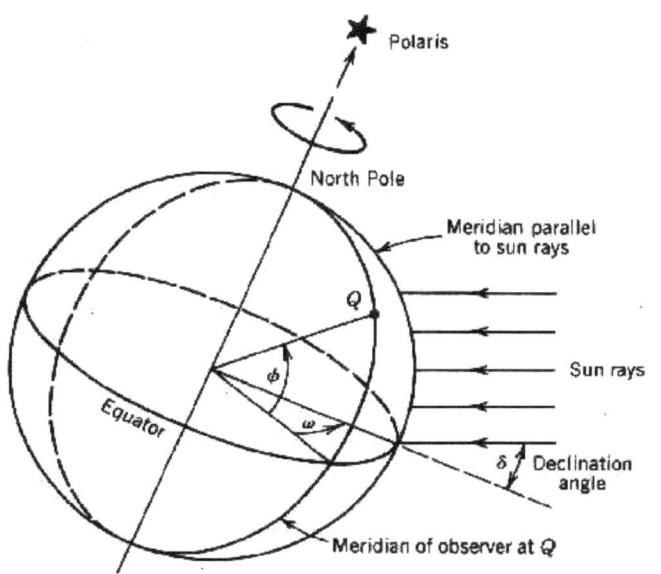

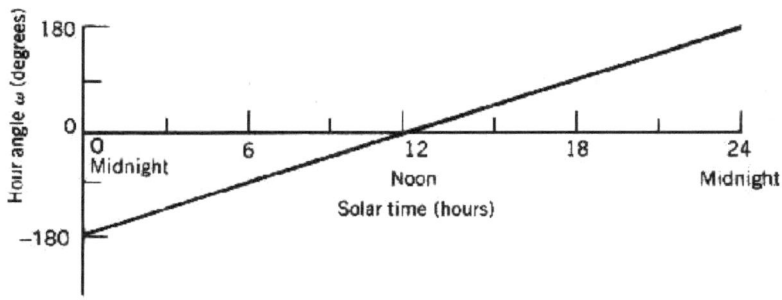

Figure - 2.29 : Hour Angle and its variations during a day.

SOLAR RADIATION GEOMETRY

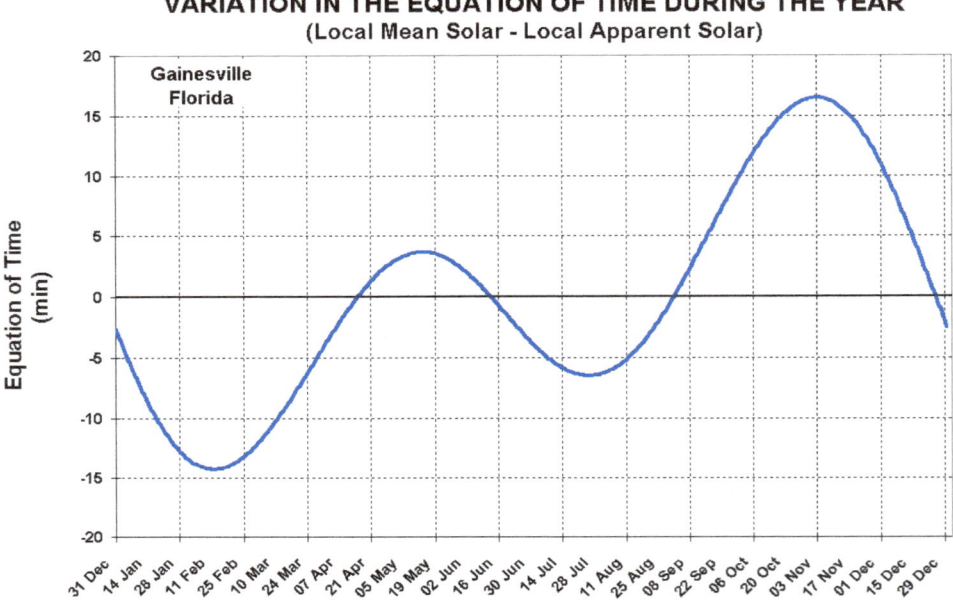

Figure – 2.30 : Equation of Time – Graph.

Problem 2 : Determine the LST and Declination at a location latitude 23°51' N, longitude 77°30' E at 12.30 IST on June 20. EOT correction = – (1' 02"). Standard Time Longitude for IST = 82.5°.

Solution :

Given Data : IST = 12:30 hrs.
Latitude, θ = 23°51' N
Longitude, ϕ = 77°30' E
Standard Time Longitude for IST = 82.5°

SOLAR RADIATION GEOMETRY

Day = June 20 => N = (31 + 28 + 31 + 30 + 31 + 20) = 171

EOT correction = - (1' 02")

LST = Standard Time ± 4 (Standard Time Longitude - Longitude of Location) + (Equation of Time Correction)

LST = 12h 30' + 4 (82° 30' - 77° 30') + (- 01' 02")

= 12h 30' 0" + 20' 0" - 01' 02"

= 12h 48' 58", i.e. **12:48:58 hrs.**

$$\delta = 23.45° \cdot \sin\left[\frac{360°}{365} \cdot (N + 284)\right]$$

δ = 23.45°. sin [(360 / 365) (171 + 284)] = **23.44°**

Problem 3 : Determine the LST and Declination at a location latitude 23°15' N, longitude 67°30' E at 02.30 IST on October 02. EOT correction = (9' 02"). Standard Time Longitude for IST = 82.5°.

Solution :

Given Data :   IST = 02:30 hrs.
Latitude, θ = 23°15' N
Longitude, φ = 67°30' E
Standard Time Longitude for IST = 82.5°
Day = October 02
=> N = (31 + 28 + 31 + 30 + 31 + 30 + 31 + 31 + 30 + 02)
= 275

EOT correction = (09' 02")

LST = Standard Time ± 4 (Standard Time Longitude – Longitude of Location) + (Equation of Time Correction)

LST = 02h 30' + 4 (82° 30' – 67° 30') + (09' 02")

= 02h 30' 0" + 60' 0" + 09' 02"

= 02h 99' 02" = 03h 39' 02" i.e. **03:39:02 hrs.**

$$\delta = 23.45° \cdot \sin\left[\frac{360°}{365} \cdot (N + 284)\right]$$

δ = 23.45°. sin [(360 / 365) (275 + 284)] = **-4.612°**

DERIVED ANGLES : Related to the **relative position of the Sun** :

Altitude Angle or Solar Altitude (α) : **Vertical angle** between the **projection of Sun's rays on the horizontal plane on Earth's surface** and **the direction of Sun's ray.**

The altitude angle α **is maximum** at **solar noon.**

Zenith Angle (θ_z) : **Complimentary angle of Solar Altitude Angle,** i.e. Vertical **angle** between **Sun's rays and a line perpendicular to the horizontal plane** though the **point,** i.e. angle between **the beam from the sun and the vertical.**

SOLAR RADIATION GEOMETRY

Solar Azimuth Angle (γ_s) : Solar Angle in degrees **along the horizon east or west of North**. It is a **horizontal angle** measured from **North to horizontal projection of sun's rays.**

It is considered +ve west-wise.

Figure 2.31 shows derived solar angles.

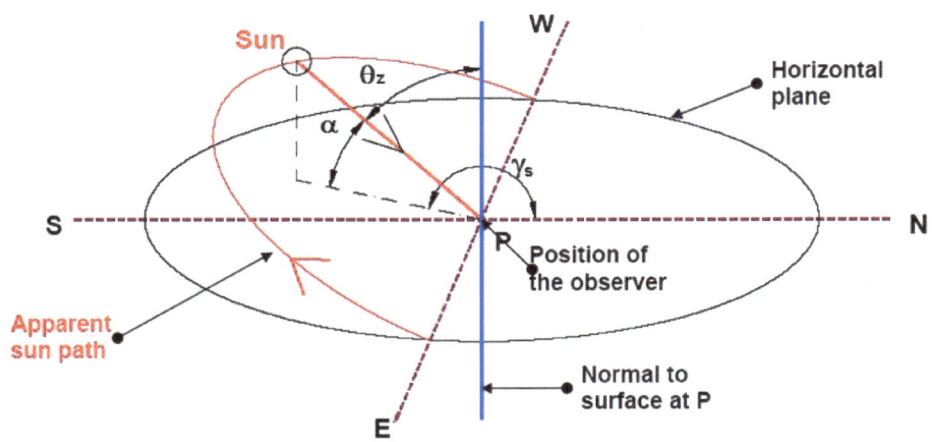

Figure – 2.31 : Derived Solar Angles.

DERIVED ANGLES : Related to the **orientation of surface intercepting solar radiation** :

Slope or Tilt angle (s) : Angle made by the **plane surface** with **the horizontal**.

SOLAR RADIATION GEOMETRY

It is : **+ve** : for surfaces slopping towards South

-ve : for surfaces slopping towards North.

Figure 2.32 shows Slope or Tilt angle.

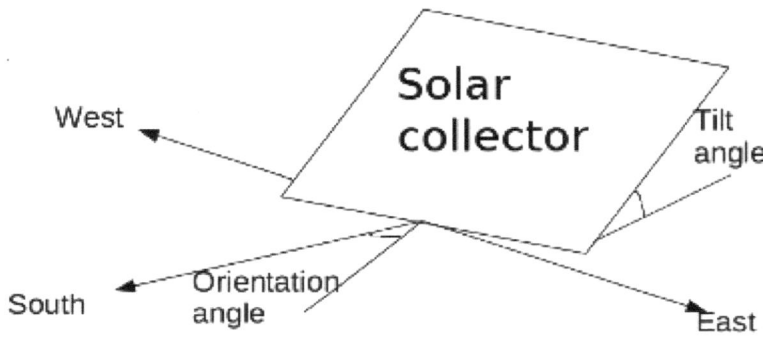

Figure – 2.32 : Slope or Tilt Angle (s).

<u>**Angle of Incidence (θ)**</u> : When Tilted Surfaces are involved, angle between **Sun rays** and **normal to surface** under consideration.

SOLAR RADIATION GEOMETRY

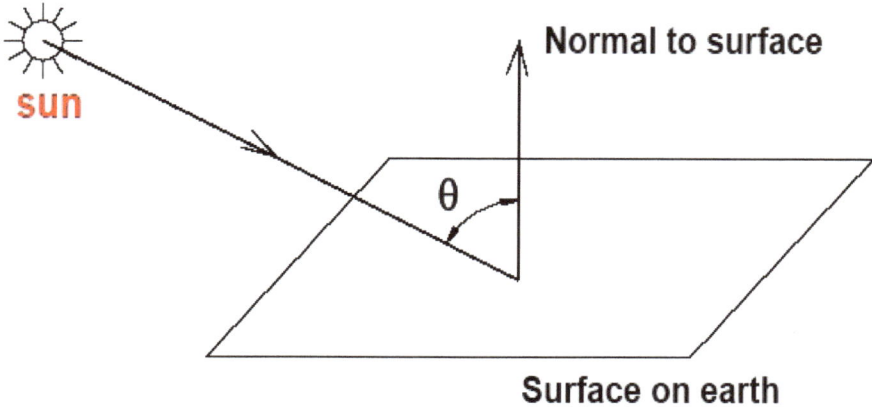

Figure – 2.33 : Angle of Incidence (θ).

Surface Azimuth Angle (γ) : For tilted surfaces, angle of deviation of the **normal** to the surface from the **local meridian**.

It is angle between the **normal to the surface and south**.

For south-facing surface, γ = 0.

For west-facing surface, γ = 90° and so on.

East-ward : +ve, West-ward : -ve.

SOLAR RADIATION GEOMETRY

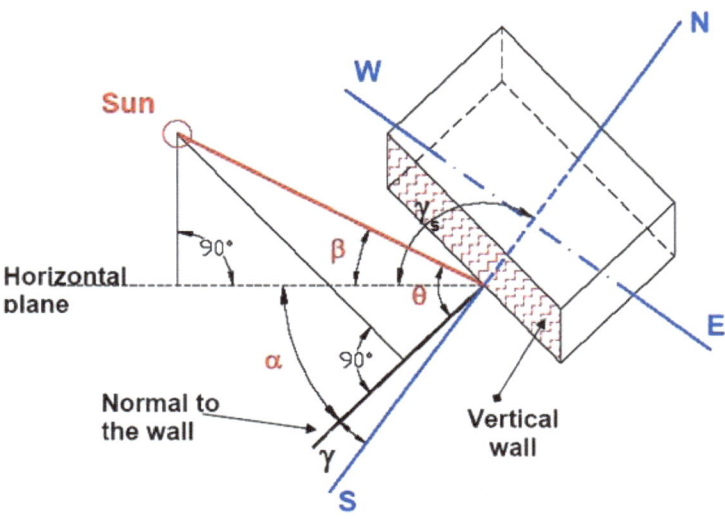

Figure – 2.34 : Surface Azimuth Angle (γ).

RELATION BETWEEN BASIC AND DERIVED ANGLES (related to relative position of the Sun) :

Basic Solar Angles :

1. Latitude of location (ϕ_l or l).

2. Declination (δ or d).

3. Hour Angle (ω or h).

Derived Solar Angles :

1. Altitude Angle (α).

2. Zenith Angle (θ_z).

3. Solar Azimuth angle (γ_s).

$\cos \theta_z = \cos \phi \cos \omega \cos \delta + \sin \phi \sin \delta$

$\qquad = \sin \alpha$ ...... (as $\theta_z = \pi/2 - \alpha$)

$\cos \gamma_s = \sec \alpha (\cos \phi \sin \delta - \cos \delta \sin \phi \cos \omega)$

$\sin \gamma_s = \sec \alpha \cos \delta \sin \omega$

GENERAL EQUATION FOR ANGLE OF INCIDENCE (θ) :

$\cos \theta = \sin \phi_l (\sin \delta \cos s + \cos \delta \cos \gamma \cos \omega \sin s) + \cos \phi_l (\cos \delta \cos \omega \cos s - \sin \delta \cos \gamma \sin s) + \cos \delta \sin g \sin \omega \sin s$ .... (i)

For Vertical Surfaces :

Slope, s = 90°

So, from Equation (i) :

$\cos \theta = \sin \phi_l (\sin \delta \cos s + \cos \delta \cos \gamma \cos \omega \sin s) + \cos \phi_l (\cos \delta \cos \omega \cos s - \sin \delta \cos \gamma \sin s) + \cos \delta \sin \gamma \sin \omega \sin s$

Putting Sin s = 1, Cos s = 0, we have

$\cos \theta = \sin \phi_l \cos d \cos \gamma \cos \omega - \cos \phi_l \sin \delta \cos \gamma + \cos \delta \sin \gamma \sin \omega$ .... (ii)

For Horizontal Surfaces :

Slope, s = 0°, Zenith Angle $\theta_z = \theta$

So, from Equation (i) :

$$\cos \theta = \sin \phi_l (\sin \delta \cos s + \cos \delta \cos \gamma \cos \omega \sin s) + \cos \phi_l (\cos \delta \cos \omega \cos s - \sin \delta \cos \gamma \sin s) + \cos \delta \sin \gamma \sin \omega \sin s$$

Putting Cos s = 1, Sin s = 0, we have

Hence, $\cos \theta = \sin \phi_l \sin d + \cos \phi_l \cos \delta \cos \omega = \sin \alpha$ ……….. **(iii)**

i.e. $\cos \theta = \cos \theta_z = \sin \alpha$ (As, $\sin \alpha = \cos \theta_z$)

For Surfaces facing due south :

Surface Azimuth Angle, $\gamma = 0°$,

So, from Equation (i) :

$$\cos \theta = \sin \phi_l (\sin \delta \cos s + \cos \delta \cos \gamma \cos \omega \sin s) + \cos \phi_l (\cos \delta \cos \omega \cos s - \sin \delta \cos \gamma \sin s) + \cos \delta \sin \gamma \sin \omega \sin s$$

Hence, putting Cos γ = 1, Sin γ = 0, we have,

$\cos \theta = \sin \phi_l (\sin \delta \cos s + \cos \delta \cos \omega \sin s) + \cos \phi_l (\cos d \cos \omega \cos s - \sin \delta \sin s)$ ……… **(iv)**

⇨ $\cos \theta = \sin \delta \sin (\phi - s) + \cos \delta \cos \omega \cos (\phi - s)$

For Vertical Surfaces facing due south :

Surface Azimuth Angle, $\gamma = 0°$, slope, s = 90°

So, from Equation (i) :

SOLAR RADIATION GEOMETRY

$\cos \theta = \sin \phi_l (\sin \delta \cos s + \cos \delta \cos \gamma \cos \omega \sin s) + \cos \phi_l (\cos \delta \cos \omega \cos s - \sin \delta \cos \gamma \sin s) + \cos \delta \sin \gamma \sin \omega \sin s$

Putting $\cos \gamma = 1$, $\sin \gamma = 0$, $\cos s = 0$, $\sin s = 1$, we have

$\cos \theta = \sin \phi_l \cos \delta \cos \omega + \cos \phi_l \sin \delta$ .... (v)

DAY LENGTH :

The day length, i.e. the duration of day varies due to changes in the declination and hence corresponding changes in the solar altitude angle for the given location.

At the time of sunrise or sunset, the Zenith angle, $\phi_z = 90°$. Substituting this in equation (iii) obtained earlier :

$\cos \theta = \sin \phi_l \sin \delta + \cos \phi_l \cos \delta \cos \omega = \sin \alpha$ (= 0)

Sun Rise Hour Angle (ω_s) :

$\cos \omega_s = -(\sin \phi \sin \delta) / (\cos \phi \cos \delta)$

$\Rightarrow \cos \omega_s = -(\tan \phi \tan \delta)$

$\Rightarrow \omega_s = \cos^{-1}(-\tan \phi \tan \delta)$ .... (vi)

Since, $\omega_s = $ **15° per hour**,

Day Length (in hours) = $t_d = (2 \omega_s / 15)$

i.e.

$t_d = (2 / 15) \cos^{-1}(-\tan \phi \tan \delta)$

Thus, **day length** is a <u>function of</u> **latitude** ϕ and **solar declination** δ.

Problem – 4 : Calculate the number of daylight hours in Delhi on 24th December 1996. Take Latitude = 28°35'N.

Solution :

Latitude, $\theta = 28°35'N = 28\frac{35}{60} = 28.583°$

Number of day, n = 366 – (31 – 24) = 359.
 (as, 1996 is a leap year.)

Declination = $\delta = 23.45 \, Sin \, [\frac{360}{365}](n + 284)$

$\quad = 23.45 \, Sin \, [\frac{360}{365}](359 + 284)$

$\quad = -23.387°$

Day length, $t_d = \frac{2}{15} cos^{-1}[-tan\delta . tan\emptyset]$

$\quad = \frac{2}{15} cos^{-1}[-tan(-23.387). tan 28.583]$

$\quad = 10.183$ hours.

About the Author :

Vanita N. Thakkar – a **Technocrat** having more than **20 years** of experience in various fields viz.

Academics,
Design,
Research and Development,
Valuation etc.

Academic Qualifications : B. E. (Mech.) and M. E. (Thermal Sc.) from M. S. University of Baroda

Experience : Industrial – 11+ years, Academic – 11+ years in MSU and GU / GTU.

Publications in National and International Journals.

Expert and Reviewer in Technical Magazines, journals, etc.

Professional Qualifications :
- IBBI approved Registered Valuer for Plant & Machinery under Section 247 of Companies Act 2013
- Registered as Valuer for Plant & Machinery Under Section 34AB of W.T.A., 1957.
- BEE Certified Energy Manager.

- MNRE approved Solar Chartered Engineer (1st lady from Gujarat to have become Solar CE).
- Research Scholar, C U Shah University, Surendranagar (Gujarat, India).

Actively associated with several organizations in the field of Technology and Art (Vanita is an All India Radio Approved Singer-Composer, Poetess-Writer).

YouTube Channel : https://www.youtube.com/channel/UC3iF2D_tVz-RpWIIA8MJsow

E-mail : vanitaa.thakkar@gmail.com, omkaarees@gmail.com

www.ingramcontent.com/pod-product-compliance
Lightning Source LLC
Chambersburg PA
CBHW040327220526
45473CB00009B/2595